To my mom and dad who always told me
I could do anything I put my mind to. - Tina

The Firehouse Gang
Text © 2010 – Tina Cook
Illustrations © 2010 - Christopher A. Starkey
Printed in United States of America (USA)
Published by Great Books 4 Kids
(http://www.greatbooks4kids.org)
All rights reserved.

ISBN 978-0-9822168-8-0

For as long as anyone could remember, Townsville had been without a fire station.

All the people of Townsville wanted a fire station with fire trucks and heroes to protect them.

The town decided to come together and vote about building a firehouse.

The votes were in...

The votes were tallied, and there was much clapping and cheering. Townsville would have a fire department at last! "But where will the firehouse be built?" thought the people of Townsville.

Freddy the furniture store smiled. He knew that he was a perfect place for a firehouse. He was the oldest and wisest building in Townsville, and he had been empty for a long time.

He shook with excitement at the thought of becoming Townsville's first fire station.

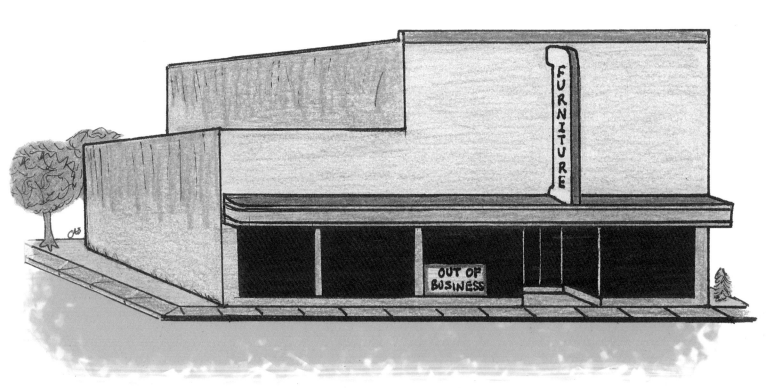

"Freddy the Firehouse, what a nice name it will be!" he thought.

The next few weeks went by very quickly for Freddy and Townsville.

Freddy had been transformed into a firehouse. He had five garage doors, a flag, a siren, and a flashing red light!

"All Townsville needs now is a few fire trucks," thought Freddy.

That very afternoon, Freddy had a visit from his old friend, Tony the Taxi.

"Hello Tony. How are you?" said Freddy happily.

"I'm great," said Tony the Taxi. " I almost didn't recognize you Freddy! You look wonderful. "

"Yes," said Freddy the Firehouse, "I am almost ready!"

"Almost ready?" asked Tony.

"We still need fire trucks!" explained Freddy.

"I have some friends you should meet," said Tony the Taxi. "I will bring them by tomorrow."

The next morning Freddy had a visit from three new friends.

"Hi!" said the big loud truck.
"My name is Tom the Tiller."

Tom was red and white with
a big ladder and lots of tools.

"I drive fire fighters when they have a fire call."

"I help get them to the top of buildings with my ladder!"

"Of course, I am the most important fire truck!"

The next fire truck was smaller than Tom and much quieter.

He was red and white with lots of shiny buttons and knobs.

"I am Edgar the Engine!" he said proudly.

"I can drive the fire fighters and carry their tools and ladders."

"I can also carry water to help fight fires."

"So you can see that I am the most important fire truck!"

Next Freddy met Bill the Brush Truck. He was a strange looking noisy fire truck. He was even smaller than Edgar and had really big tires.

Bill was red and white and had a fire hose and a water tank.

He had some tools, but did not look like anything Freddy had ever seen.

"Hello! My name is Bill the Brush Truck. I can drive fire fighters anywhere to get to a fire!"

"I can go up hills and through the mud. I can go where those other two fire trucks cannot!"

"I guess that makes me the most important fire truck!"

"Well, you all seem to be very important in your own ways," said Freddy the Firehouse. "I believe that Townsville could use all three of you!"

Tom, Bill and Edgar cheered loudly!

Freddy the Firehouse gave each of the fire trucks a place of their very own to park. All the trucks were excited to have a new home. Now Townsville had a working Firehouse. Hooray for the Townsville's fire heroes!

"Now that you are all settled in," said Freddy, "it is time for our first fire safety lesson."

"Let's learn about how people call us for help in an emergency," said Freddy.

Tom, Edgar and Bill paid close attention to Freddy. They knew fire safety was a very important lesson!

"When someone needs help in an emergency they call a special phone number."

"The person who answers the call for help will ask for important information. They may ask for your name, your phone number, the address where you are, and what kind of help you need."

"What else should people do when they call to report an emergency?" asked Freddy.

"You should never call an emergency number as a joke," said Bill the Brush Truck.

"Remain calm on the phone," added Edgar the Engine.

"Never hang up until the person on the other end says you can!" explained Tom the Tiller.

"Excellent!" said Freddy. "That is enough for one day. Lets get some rest. We have another busy day tomorrow!"

"Good night Edgar, good night Tom, good night Bill," said Freddy the Firehouse.

The citizens of Townsville slept peacefully that night. They knew which numbers to call in an emergency, and they knew that they had a Firehouse full of heroes to protect them.

SAFETY LESSON

What phone number would you call in these emergencies?

```
IN CASE OF EMERGENCY
         DIAL
  🔥 FIRE   _____
  🛡 POLICE _____
  ✳ EMS    _____
  MY ADDRESS _____
  _____
  MY PHONE # _____
```

They might all be the same phone number or they might be different numbers!

SAFETY IN ACTION

Practice dialing your emergency numbers.

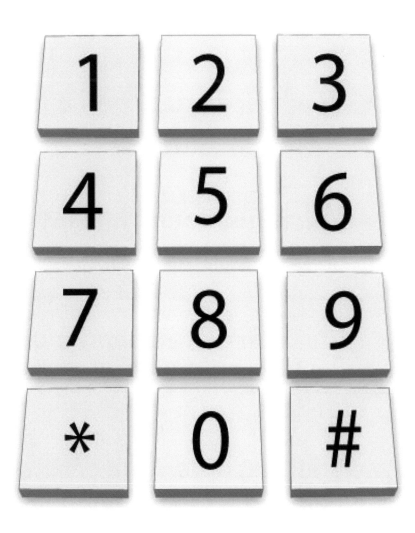

MY FIREHOUSE GANG QUIZ

Question 1)   What was the name of Townsville's new Fire Station?

Question 2)   Which fire truck had a long ladder?

Question 3)   Which fire truck has lots of shiny buttons and knobs?

Question 4)   Which fire truck had big tires and can drive fire fighters anywhere to get to a fire?

Question 5)   What important information will you need when you call for help in an emergency?

Answer 1)   Freddy the Firehouse.

Answer 2)   Tom the Tiller.

Answer 3)   Edgar the Engine.

Answer 4)   Bill the Brush Truck.

Answer 5)   Your name, your phone number, the address of where you are at, and what kind of help you need.

LaVergne, TN USA
13 February 2011
216340LV00003B